PowerKids Readers:
My World of Science™

Bilingual Edition
English/Spanish
Edición bilingüe

Electricity in My World

La electricidad en mi mundo

Joanne Randolph

Traducción al español: María Cristina Brusca

The Rosen Publishing Group's
PowerKids Press™ & **Editorial Buenas Letras**™
New York

For Linda Lou and Lucas

Published in 2006 by The Rosen Publishing Group, Inc.
29 East 21st Street, New York, NY 10010

Copyright © 2006 by The Rosen Publishing Group, Inc.

First Edition

Photo Credits: Cover (outlet), p.22 (balloons) © Royalty-Free/Corbis; Cover (glass sphere) © David Langfield; Eye Ubiquitous/Corbis; Cover (girl), p. 19 © David Young-Wolff/Getty Images; p. 5 @ BrandXPictures/Getty Images; p. 7 © Wartenberg/Picture Press/Corbis; pp. 9, 22 (power plant) © Galen Rowell/Corbis; p. 11 © Roger Ball/Corbis; p. 13 © Morton Beebe/Corbis; p. 15 © Paul A. Souders/Corbis; p. 17 © Bob Krist/Corbis; p. 21 © Reed Kaestner/Corbis; p. 22 (street lamp) © Justin Hutchinson/Corbis; p. 22 (wires) © Cooperphoto/Corbis.

Library of Congress Cataloging-in-Publication Data

Randolph, Joanne.
[Electricity in my world. Spanish & English] Electricity in my world = La electricidad en mi mundo / Joanne Randolph ; traducción al español, María Cristina Brusca.— 1st ed.
p. cm. — (My world of science)
Includes bibliographical references and index. ISBN 1-4042-3319-9 (lib. bdg.)
1. Electricity—Juvenile literature. I. Title: Electricidad en mi mundo. II. Title.
QC527.2.R3618 2006
537—dc22
2005007159

Manufactured in the United States of America

Contents

Contenido

Electricity is an important part of our world. We need electricity to turn on the lights. We need it to make toast for breakfast.

La electricidad es una parte importante de nuestro mundo. Necesitamos de la electricidad para encender las lámparas. La necesitamos para preparar las tostadas de nuestro desayuno.

Electricity is a kind of energy. Energy is the ability to do work. Energy from the Sun heats Earth. When we eat food it gives our bodies energy.

La electricidad es una forma de energía. La energía es la capacidad de hacer un trabajo. La energía del Sol calienta la Tierra. Cuando comemos, nuestros cuerpos reciben energía de los alimentos.

7

We use the energy from electricity to power many things. It comes to our homes from a power plant. It travels through wires to reach our homes.

Usamos la energía de la electricidad para hacer funcionar muchas cosas. La electricidad viene a nuestros hogares desde una planta eléctrica. Para llegar a nuestras casas viaja a través de cables.

Electricity moves in a current. Wires carry the current to the places where we need it. Wires are often made of metal. Metal allows electricity to pass through easily.

La electricidad se mueve en una corriente. Los cables llevan la corriente a los lugares donde la necesitamos. Muchas veces los cables están hechos de metal. El metal le permite a la electricidad moverse con facilidad.

11

Lightning is electricity. Have you ever seen a thunderstorm? The storm charges the air. Lightning is the energy being released from the air.

Los rayos son electricidad. ¿Has visto alguna vez una tormenta eléctrica? La tormenta carga el aire de energía. Los rayos se producen cuando la energía se descarga desde el aire.

Static electricity is another kind of electricity. Have you ever noticed what happens to your hair if you rub it on a cold day? Static electricity makes the hair stand out from your head.

La electricidad estática es otra forma de electricidad. ¿Has notado lo que le sucede a tu cabello cuando lo frotas en un día frío? La energía estática hace que tus cabellos se paren.

The lights in your house are run by electricity. The street lamps outside are run by electricity, too. Think about how much electricity is being used to power the lights of a whole town!

Las bombillas de tu casa funcionan con electricidad. Las lámparas de la calle también funcionan con electricidad. ¡Piensa en cuánta electricidad se usa para hacer funcionar las lámparas de una ciudad!

Batteries give us electricity, too. The battery stores energy. We can put the battery in a toy or radio. The energy is turned into electricity.

Las pilas también nos dan electricidad. Las pilas guardan energía. Podemos poner pilas en un juguete o en la radio. La energía se convierte en electricidad.

Can you think of all the things around you that use electricity? Look at this picture. Can you name all the things that need electricity to run?

¿Puedes pensar en todas las cosas a tu alrededor que usan electricidad? Mira esta foto, ¿puedes nombrar las cosas que necesitan electricidad para funcionar?

Words to Know
Palabras que debes saber

battery
pila

power plant
planta eléctrica

street lamp
lámpara
de la calle

wires
cables

Here are more books to read about electricity:
Otros libros que puedes leer sobre la electricidad:

In English/En inglés
Where Does Electricity Come From?
The Clever Calvin Series
by C. Vance Cast and Sue Wilkinson
Barron's Educational

In Spanish/En español
¡Es eléctrico!
by Greg Roza
Rosen Real Readers en español

Web Sites/En Internet
Due to the changing nature of Internet links,
PowerKids Press and Editorial Buenas Letras have
developed an online list of Web sites related to the
subject of this book. This site is updated regularly.
Please use this link to access the list:

www.powerkidslinks.com/mws/electricity/

Index

Índice

Word Count: 251   Número de palabras: 269

Note to Librarians, Teachers, and Parents

PowerKids Readers are specially designed to help emergent and beginning readers build their skills in reading for information. Sentences are short and simple, employing a basic vocabulary of sight words, simple vocabulary, and basic concepts, as well as new words that describe objects or processes that relate to the topic. Large type, clean design, and stunning photographs corresponding directly to the text all help children to decipher meaning. Features such as a contents page, picture glossary, and index introduce children to the basic elements of a book, which they will encounter in their future reading experiences.